William Munk

Euthanasia

Medical Treatment in Aid of an Easy Death

William Munk

Euthanasia
Medical Treatment in Aid of an Easy Death

ISBN/EAN: 9783744760263

Printed in Europe, USA, Canada, Australia, Japan

Cover: Foto ©berggeist007 / pixelio.de

More available books at **www.hansebooks.com**

EUTHANASIA:

OR,

MEDICAL TREATMENT IN AID OF AN EASY DEATH.

BY

WILLIAM MUNK, M.D., F.S.A.

FELLOW AND LATE SENIOR CENSOR OF THE ROYAL COLLEGE OF
PHYSICIANS, &C., &C., &C.

LONDON:

LONGMANS, GREEN, AND CO.

AND NEW YORK: 15, EAST 16th STREET.

1887.

PREFACE.

——◆◆——

MUCH has been ably written on Death, and on the physiology of the various modes of Dying, by Bichat, Alison, Wilson-Philip, Symonds, and others; while but little has been written on the medical management of the Dying; or on the Euthanasia, to which such management should contribute. A short but valuable essay of less than twenty pages, "On the Treatment of the Dying", by Dr. Ferriar, of Manchester, in 1798: and a very elegant academical oration, of about the same length, at my own alma mater, Leyden, in 1794, by Professor Paradys, "Oratio de Ευθανασία Naturali; et quid ad eam conciliandam Medicina valeat":—comprise all I know that has been written *specially* on these subjects in modern times.

Sir Henry Halford, who was con-

fessedly a master in all that concerns
the management of the Dying, did much
by his example and counsel to com-
mend the subject to the attention of his
medical brethren.   But the generation he
personally influenced has passed away. His
little volume of " Essays and Orations "
contains much on this subject that is
very valuable, and not to be found else-
where.   But his remarks are unconnected ;
they occur incidentally in the course of
his various essays, and are now but
little known.   They were the result of an
experience so large, and so carefully
thought out, that I have been glad to
adduce them, whenever I could, in support
of, or in addition to, what I have had
myself to state, on the delicate and difficult
subjects considered in the following pages.

40, Finsbury Square.
Sept. 24, 1887.

CONTENTS.

I.

I.

SOME OF THE PHENOMENA OF DYING.

"Quod ad nos
Pertinet, et nescire malum est, agitamus."

HORACE.

SOME OF THE PHENOMENA
OF DYING.

———◆◆———

ONE of the wisest of our countrymen, Lord Verulam, saw reason to censure the physicians of his own time for not making the Euthanasia a part of their studies.[1]   And, although more than two

---

[1] " At nostris temporibus, Medici quasi religio est, ægrotis, postquam deplorati sint, assidere; ubi meo judicio, si officio suo, atque adeo humanitati ipsi deesse nolint, et artem ediscere et diligentiam præstare deberent, qua animam agentes facilius et mitius è vita demigrent—Hanc autem partem, inquisitionem de *Euthanasia* exteriori (ad dif-ferentiam ejus euthanasiæ quæ animæ præpara-

centuries have since elapsed, it may be
doubted whether as much attention is even
yet given to the subject as might be done,
to the obvious benefit and comfort of the
dying.

There is little to be found in medical writ-
ings on the management of the dying, or on
the treatment best adapted to the relief of
the sufferings incident to that condition. The
subject is not specially taught in any of our
medical schools ; and the young physician
entering on the active duties of his office
has to learn for himself, as best he may,
what to do, and what not to do, in the
most solemn and delicate position in which
he can be placed,—in attendance on the
dying, and administering the resources of
the medical art, in aid of an easy, gentle,
and placid death.   The whole subject of

tionem   respicit) appellamus ;   eamque inter
desiderata reponimus."   (Verulamus, De Augmentis Scientiarum, lib. iv. cap. ij.)

the Euthanasia,[1] or of a calm and easy ✓
death, in so far as it respects the physician
is in need of special study ; and of a sys-
tematic treatment that has not hitherto been
accorded to it.[2]   In the following pages I
can but trace the outlines of this subject,
leaving to abler hands that fuller treatment
which its interest and importance claim for
it.

Lord Verulam held it to be as much the
duty of the physician to smooth the bed of
death, and render the departure from this

---

[1] " *Εὐθανασία* naturalis nobis dicitur facilis
et quam minimo cum cruciatu e vita exitus, qua
tenus moriendi facilitas e causis naturalibus
proxime pendet." "Ad medicinam hujus
εὐθανασίας contemplatio pertinet : est enim
naturalis, non moralis, nisi qua tenus hæc ad illam
momenti habet plurimum.   Exteriorem idcirco
Verulamius appellavit." (Nicolai Paradysii,
Opuscula Academica, 8vo, Lugd. Batav, 1813.
Oratio de *Εὐθανασία* naturali et quid ad eam
conciliandam Medicina valeat, pp. 63 et 65.)

[2] " A medicis vix inchoatum, nedum pertracta-
tum huc usque esset." (Paradysius, p. 63.)

life easy and gentle, as it is to cure diseases and restore health.[1]   And this doctrine, so accordant with the best principles of our nature,[2] is commended to us by that most estimable and judicious of modern physicians, Dr. Heberden;[3] as it was also by

---

[1] "Etiam plane censeo ad officium medici pertinere, non tantum ut sanitatem restituat: verum etiam ut dolores et cruciatus morborum mitiget: neque id ipsum solummodo, cum illa mitigatio doloris, veluti symptomatis periculosi, ad convalescentiam faciat et conducat: imo vero cum abjecta prorsus omni sanitatis spe, excessum tantum præbeat e vita magis lenem et placidum. Siquidem non parva est felicitatis pars, illa Euthanasia." (De Augmentis Scientiarum.)

[2] Sir Henry Halford, Essays and Orations read and delivered at the Royal College of Physicians. Third edition, 12mo, London, 1842. p. 84.

[3] "Magnus ille veræ philosophiæ instaurator Verulamius, queritur studium Euthanasiæ medicis haud satis cultum fuisse. Medici profecto munus est ægrotis sanitatem reddere; cum tamen ex lege naturæ erit tandem unicuique mortalium ægrotatio nulla arte medicabilis, benevolæ hujus

the example and counsel of one of the most popular and successful physicians of the present century—the late Sir Henry Halford.[1]

The process by which death is brought about varies greatly in different instances, and this according to the disease, or the organ of the body, from which it essentially results. On these diverse modes of dying, and of death, modern science has thrown much light; and with the consolatory result of showing that the process of dying, and the very act of death, is but rarely and exceptionally attended by those severe bodily sufferings, which in

---

artis professoribus conveniret, mortem inevitabilem, quantum fieri potest, terrore omni spoliare; et ubi non datum est prædam morti extorquere, sed vita necessario amittenda est, operam saltem dare, ut cum minima crudelitatis specie amittatur." (Heberdeni Gulielmi, Commentaria de Morborum Historia et Curatione. Cap De Ileo.)

[1] Essays and Orations, *ut supra passim*.

popular belief are all but inseparable from it, and are expressed and emphasized in the terms "mortal agony" and "death struggle."

Montaigne was one of the first among modern writers to oppose, by close argument, the general opinion of the painfulness of death; and he was followed in the last century with more eloquence, if with less argument, by Buffon.[1] "There is hardly any subject," writes an amiable physician, "on which books afford us more impressive topics, than the consideration of death; and perhaps there is none less studied in its intimate details. . . . It might be expected that a scene through which we must all pass should excite a closer attention especially as *the physical*

---

[1] John Ferriar, M.D., Medical Histories and Reflections. 8vo, London, 1798. Vol. iii. p. 196.

process of death loses much of its horror on a near view." [1]

Physicians, the clergy, and intelligent nurses—all, indeed, who are practically conversant with the dying—testify to the truth of this statement. Sir Henry Halford, towards the close of his medical career, and after opportunities of observation, such as have fallen to the lot of few physicians, expressed his surprise that of the great number to whom it had been his professional duty to have administered in the last hours of their lives, so few exhibited signs of severe suffering. Sir Benjamin Brodie, whose experience of death from surgical disease was second to none, states that, according to his observation, the mere act of dying is seldom, in any sense of the word, a very painful

---

[1] John Ferriar, M.D., On the Treatment of the Dying, *ut supra*, p. 191.

2 *

process.[1]   And another distinguished sur-
geon, Mr. Savory, writing on the same
subject, says, " Whatever may have been
the amount of *previous* suffering, we may
fairly assume that, except in extreme cases,
the actual process of dying is not one of
intense agony, or indeed, for the most part,
even of pain.[2] Lastly, the great anatomist,
Dr. William Hunter, bore his own dying
testimony to the same effect. He retained
his consciousness to the last, and just
before he died he whispered to his friend,
Dr. Combe, " If I had strength enough to
hold a pen, I would write how easy and
pleasant a thing it is to die." [3]

---

[1] The Works of Sir Benjamin Collins
Brodie.   Arranged by Charles Hawkins.
3 Vols., 8vo, London, 1865. Vol. i. p. 184.
[2] On Life and Death. 8vo. London,
1863, p. 175.
[3] " Ipsæ animæ discessus a corpore fit, sine
dolore, et fit plerumque sine sensu, *nonnunquam
etiam cum voluptate.*" (Vopisci   Fortunati

But of far greater weight than the observations and conclusions of medical men, however eminent, towards the determination of such a question, is the evidence of those who have been restored from the state of apparent death from drowning—a state which differs only from actual death in the possibility of reanimation under the influence of external treatment. And although the accounts given after recovery from drowning vary much, there are a number of well-attested cases which show, that in them at any rate, the loss of sensibility and conciousness has been painless, or at most attended with a feeling of oppression across the chest. The process of recovery, however, is often one of great bodily suffering.

Lastly, there are those specially interesting cases of recovery from the apparent

Plempii. de Togatorum Valetudine tuenda Commentatio. 4to. Bruxellis, 1670. p. 26.)

death of drowning, in which, although the mind has been keenly alive and active throughout, there was an entire absence of pain or other bodily suffering of any kind. The best authenticated of these instructive and suggestive instances is that of Admiral Beaufort, as described by himself in a letter to Dr. Wollaston.[1]   When a youngster on board one of H.M. ships in Portsmouth harbour, he fell into the water, and, being unable to swim, was soon exhausted by his struggles, and before relief reached him, he had sunk below the surface. All hope had fled, all exertion ceased, and he felt that he was drowning. " From the moment that all exertion had ceased," writes the admiral, "a calm feeling of the most perfect tranquillity superseded the previous tumultuous sensations—it might be called apathy, certainly not resignation, for drown-

---

[1] Autobiographical Memoir of Sir John Barrow, Bart. 8vo, London, 1847, p. 398.

ing no longer appeared to be an evil. I
no longer thought of being rescued, *nor
was I in any bodily pain. On the contrary,
my sensations were now of rather a plea-
surable cast, partaking of that dull, but
contented sort of feeling which precedes the
sleep produced by fatigue.* Though the
senses were thus deadened, not so the
mind; its activity seemed to be invigorated
in a ratio which defies all description—for
thought rose after thought with a rapidity
of succession, that is not only indescribable,
but probably inconceivable, by any one
who has not himself been in a similar
situation. The course of these thoughts
I can even now in a great measure retrace,
—the event which had just taken place,
the awkwardness that had produced it, the
bustle it must have occasioned, the effect
it would have on a most affectionate
father, and a thousand other circumstances
minutely associated with home were the
first series of reflections that occurred.

They then took a wider range—our last
cruise, a former voyage and shipwreck, my
school, the progress I had made there and the
time I had misspent, and even all my boyish
pursuits and adventures.   Thus travelling
backwards, every past incident of my life
seemed to glance across my recollection in
retrograde succession; not, however, in
mere outline as here stated, but the picture
filled up with every minute and collateral
feature; in short, the whole period of my
existence seemed to be placed before me in
a kind of panoramic review, and each act
of it seemed to be accompanied by a consciousness of right or wrong, or by some
reflection on its cause or its consequences;
indeed, many trifling events which had
been long forgotten, then crowded into my
imagination, and with the character of recent
familiarity."   Certainly two minutes did
not elapse from the moment of suffocation
to that of being hauled up; and according
to the account of the lookers on, he was

very quickly restored to animation. "My feelings," continues Admiral Beaufort, "while life was returning, were the very reverse in every point of those which have been described above. One single but confused idea —a miserable belief that I was drowning dwelt upon my mind, instead of the multitude of clear and definite ideas which had recently rushed through it—a helpless anxiety—a kind of continuous nightmare seemed to press heavily on every sense, and to prevent the formation of any one distinct thought, and it was with difficulty that I became convinced that I was really alive. Again, *instead of being absolutely free from all bodily pain, as in my drowning state*, I was now tortured by pain all over me."

I have given this case at some length, because it seems to throw a new light on the act of dying, and because analogous instances are probably not uncommon. Admiral Beaufort tells us that he had heard from two or three persons, who had recovered from a similar state, a detail of their feelings, which resembled his own as nearly as was consistent with their different constitutions and dispositions. Sir Benjamin Brodie mentions an instance in a sailor; [1] De Quincey records a like instance in a female, a near relative of his own; [2] and

---

[1] "A sailor who had been snatched from the waves, after lying for some time insensible on the deck of the vessel, proclaimed on his recovery that he had been in Heaven, and complained bitterly of his being restored to life as a great hardship. The man had been regarded as a worthless fellow; but from the time of the accident having occurred, his moral character was altered, and he became one of the best conducted sailors in the ship" (The Works of Sir Benjamin Brodie, vol. i. p. 184.)

[2] I was once told by a near relative of mine —says De Quincey—that having in her childhood fallen into a river, and being on the very verge of death but for the assistance which reached her at the last critical moment, she saw in a moment her whole life, clothed in its forgotten incidents, arrayed before her as in a mirror, not successively, but simultaneously;

I have myself heard of two similar cases, but the details are not sufficiently precise to justify their narration here.

---

and she had a faculty developed as suddenly for comprehending the whole and every part. The heroine of this remarkable case, continues De Quincey, was a girl about nine years old ; and there can be little doubt that she looked down as far within the *crater* of death—that awful volcano—as any human being ever *can* have done that has lived to draw back and to report her experience. Not less than ninety years did she survive this memorable escape, and I may describe her as in all respects a woman of remarkable and interesting qualities. She enjoyed throughout her long life serene and cloudless health; had a masculine understanding; reverenced truth not less than did the Evangelists; and led a life of saintly devotion, such as might have glorified Hilarion or Paul! I mention these traits as characterising her in a memorable extent, that the reader may not suppose himself relying upon a dealer in exaggerations, upon a credulous enthusiast, or upon a careless wielder of language. Forty-five years had intervened between the first time and the last time of her telling me this anecdote, and not one iota had

In fact, all the best and all the most direct evidence that the subject admits of,

---

shifted its ground amongst the incidents, nor had any of the most trivial of the circumstances suffered change. How long the child lay in the water was probably never inquired earnestly until the answer had become irrecoverable : for a servant to whose care the child was then confided, had a natural interest in suppressing the whole case. From the child's own account it would seem that asphyxia must have announced its commencement. A process of struggle and deadly suffocation was passed through half-consciously. This process terminated in a sudden blow apparently *on* or *in* the brain, after which there was no pain or conflict: but in an instant succeeded a dazzling rush of light; immediately after which came the solemn apocalypse of the entire past life. (De Quincey's Works, Edinb., 1862, Vol. I., Confessions of an English Opium-Eater, p. 259.) Sir Dyce Duckworth reminds us that the mental condition of some who have been put to sleep with anæsthetics may throw some light on this matter. " Patients," says he, " have told us they dreamed they were transported from earth and carried off into space, were supremely happy and at rest : but that on goes to show, that as a rule, the immediate act of dying is in no sense a process of severe bodily suffering—or, indeed, for the most part even of pain.

The common belief that the act of dying is one of severe bodily suffering is due probably in part to theoretical views of the nature of the event itself; [1] but, principally, to the occurrence of conditions, physiological or pathological, which precede or accompany that act, and the nature and import of which are misinterpreted. Doubtless also, it is due in no small degree to confounding the actual stage of dying, with those urgent

---

gradually recovering consciousness, they seemed to light back again upon this world, were most reluctant to leave the Elysium they had reached, and to recommence their earthly toils and struggles " (The Agony of Dying, in Monthly Paper of the Guild of St. Barnabas for Nurses. Vol. iii. p. 81).

[1] J. A. Symonds, M.D., Art. Death, in the Cyclopædia of Anatomy and Physiology, 4 vols., royal 8vo, Lond.   Vol. i. p. 800.

symptoms of disease that precede and lead up to it, and which are often as severe or more so in those who are to recover, as in those who are to die. As a rule, to which there are doubtless exceptions, the urgent symptoms of disease subside, when the act of dying really begins. "A pause in nature, as it were, seems to take place, the disease has done its worst, all strong action has ceased, the frame is fatigued by its efforts to sustain itself, and a general tranquillity pervades the whole system." [1]

Again, convulsions, which so often attend the process of dying, are accepted in evidence of suffering, when in fact they are the reverse, for they imply a loss of consciousness and sensibility, and therefore, of the capacity to feel pain. They are automatic, and in all essential respects like the convulsions of epilepsy, of which the subject is wholly unconscious. The convulsive

[1] Halford, p. 18.

movements that sometimes attend the last
moments, and with which the person expires,
constituting the so-called "death struggle,"
are doubtless of the same painless character.

Some few, however, do really suffer
grievously in dying, and expire in great
bodily torture.   This   occurs   in   some
diseases of the heart and great vessels of
the chest, in angina pectoris, and in ileus.
But especially in that most fearful of
diseases, hydrophobia, in tetanus, and in
spasmodic cholera—in maladies charac-
terized by spasm of the external muscles,
as distinguished from their convulsion, for
spasm implies no such unconsciousness as
does convulsion, but the reverse.   Such
cases are rare, but they are so terrible that
they fix themselves in the memory, exert
an undue influence on the judgment, and,
although really exceptional in occurrence,
and in the sufferings they entail, come to
be regarded as but extreme instances of
what is assumed to be the universal and

inevitable lot of the dying. Happily for mankind it is not so.

So long as consciousness and intelligence continue, and they often do so to the last, the influence of mind and of the emotions on the bodily process of dying must be kept steadily in view. They are well-nigh as potential in the dying man as they are in the healthy. Hope is as soothing and fear as depressing in the one condition as in the other. To the dying· there is no greater solace and cordial than hope—it is the most soothing and cheering of our feelings, and if, when all hope of life and in the present has fled, the dying man can dwell with hope and confidence upon his future, it will be well for him. The retrospect of a well-spent life, " memoria bene actæ vitæ, multorumque benefactorum recordatio" is a cordial of infinitely more efficacy than all the resources of the medical art ;[1] but a firm belief in the mercy of God,

---

[1] Halford, p. 14.

and in the promises of salvation will do
more than anything in aid of an easy, calm,
and collected death.   To those who are
sceptical on this point, and such there are,
I would remark, that unless a man has himself felt the influence of religion on his own
mind, he is unable fully and accurately to
understand its influence on others.   If I
may trust my own experience I should say,
that in the aggressive *dis*believer, as in the
mere passive agnostic, doubt and anxiety
as to his future is all but sure to obtrude
itself on his last conscious moments, disturb
them, and render such an euthanasia as we
contemplate, impossible.

" The less fear a reasonable man entertains
of death," says Zimmerman, " the more
placid is he in his last moments."   Happily
such dread or terror of death as disturbs
the dying is rare.   For the most part an
urgent fear of death, when it does exist, is
observed not so much at the moment when
death is actually impending, as it is at that

earlier period when the individual realizes
for the first time that he is about to die.
The shock at *that* moment may be great,
but it is for the most part transient, and
" the subsequent contemplation of approaching death seems to be far less
terrible." [1]   A torpor seems indeed to
steal softly over the whole being as death
approaches, and the earnestness to live
abates, as the possession of life, from whatever cause, is gradually withdrawn.   Sir
Henry Halford tells us that of the great
number to whom he had administered in
the last hours of their lives, he had felt
surprised that so few have appeared reluctant to go to " the undiscovered country
from whose bourn no traveller returns." [2]

No one, writes Mr. Savory, who has often
stood at the bedside of the dying, " can have
failed to be struck by the fact of the com-
parative or complete absence of dismay as

---

[1] Savory, *ut supra*, p. 178.   [2] p. 74.

death draws near. Often, no doubt, the
mind is otherwise too fully occupied, . . .
but even in the absence of this and all dis-
tracting influences, and with a clear conviction that the approaching change is near
at hand,—the mind is calm and collected,
the thoughts serene, there is no quailing,
no giving way." [1]

The nature of the disease under which a
person succumbs, would seem to exert some
influence in this respect. Sir Benjamin
Brodie says, " I have myself never known
but two instances, in which, in the act of
dying there were manifest indications of
the fear of death. The individuals to
whom I allude were unexpectedly destroyed
by hemorrhage, which from peculiar circumstances, it was impossible to suppress.
The depressing effects which the gradual
loss of blood produced on their corporeal

---

[1] On Life and Death, 8vo, London, 1863,
p. 177.

system seemed to influence their minds, and
they died earnestly imploring the relief
which art was unable to afford." [1]

When the intimation that death is at
hand has been postponed to the latest possible moment, it comes upon the sufferer
so late, that there may not be time for him
to get over the shock of the first impres-
sion, and regain his serenity. Alarm associates itself with the act of dissolution,
which is imminent, or has already commenced, disturbs its even, easy, tenor, and
explains some at least of the harrowing
scenes that occasionally mark the death-
bed. An earlier intimation [2] to the dying

---

[1] Brodie *ut supra*, p. 185.

[2] " I think there is reason for affirming that
the risk of evil from this cause is rated generally
above the truth. In cases of imminent danger,
the mind is not always, or even commonly, to
be interpreted by the rule of health. Mental
emotions are often altered in kind, or greatly
abated in degree. Death itself is beheld under
different views—a fact familiar to all who have

person of the great change he is about to
undergo is in all respects desirable, and if
the communication be made tenderly and
with prudence, nothing but good is likely
to result from it.   An important question
here presents itself.   By whom should that
communication be made?

"You will forgive me," said Sir Henry
Halford at one of the evening meetings at
the College of Physicians, " if I presume to
state what appears to me to be the conduct
proper to be observed by a physician in
withholding, or making his patient ac-
quainted with, his opinion of the probable
issue of a malady manifesting mortal symptoms.   I own I think it my first duty to

---

watched over these scenes, and regarded the patient apart from those who are grieving around his
death-bed.   Suspicion of a painful truth often
disturbs much more than the truth plainly
stated." (Sir Henry Holland's Medical Notes
and Reflections.   Third edition, 8vo.   Lond.
1853, p. 362).

protract his life by all practicable means,
and to interpose myself between him and
everything which may possibly aggravate
his danger. And unless I shall have found
him averse from doing what was necessary
in aid of my remedies, from a want of a
proper sense of his perilous situation, I
forbear to step out of the bounds of my
province in order to offer any advice which
is not necessary to promote his cure. At
the same time, I think it indispensable to
let his friends know the danger of his case
the instant I discover it. An arrangement
of his worldly affairs, in which the comfort
or unhappiness of those who are to come
after him is involved, may be necessary;
and a suggestion of his danger by which
the accomplishment of this object is to be
obtained, naturally induces a contemplation
of his more important spiritual concerns, a
careful review of his past life, and such
sincere sorrow and contrition for what he
has done amiss, as justifies our humble

hope of his pardon and acceptance here-
after. If friends can do their good offices
at a proper time, and under the suggestions
of the physician, it is far better that they
should undertake them than the medical
adviser. They do so without destroying
his hopes, for the patient will still believe
that he has an appeal to his physician,
beyond their fears; whereas, if the physician lay open his danger to him, however
delicately he may do this, he runs a risk
of appearing to pronounce a sentence of
condemnation to death, against which there
is no appeal, no hope. . . . But friends
may be absent, and nobody near the patient
in his extremity, of sufficient influence or
pretension to inform him of his dangerous
condition. And surely it is lamentable to
think that any human being should leave
the world unprepared to meet his Creator
and Judge, 'with all his crimes broad
blown.' Rather than so, I have departed
from my strict professional duty, and have

done that which I would have done by
myself, and have apprized my patient of
the great change he was about to undergo.

"In short, no rule, not to be infringed
sometimes, can be laid down on this subject. Every case requires its own con-
siderations; but you may be assured, that
if good sense and good feeling be not
wanting, no difficulty can occur which you
will not be able to surmount with satisfaction to your patient, his friends, and your-
selves." [1]

In some instances the patient himself is
the first to discover, and this from his own
internal feelings, that he is about to die,
and he announces the fact calmly, and for
the most part without alarm, to those about
him.

Although a fear of death in itself, or for
one's own sake, is rare and exceptional, the
last moments of too many are made miser-

---

[1] Halford, p. 76.

able by solicitude for those they will leave
behind, and their end is often one of great
mental anguish. " Such have clung to life
anxiously, painfully, but they were not
influenced so much by a love of life for
its own sake, as by the distressing prospect
of leaving children, dependent upon them,
to the mercy of the world, deprived of
their parental care." [1]

In some dying persons consciousness and
the intellect remain perfect to the last.
The cases in which this is observed will
be found to agree in the fact that the brain
is correspondently unimpaired; they are
for the most part chronic diseases of the
chest and abdomen. If the character of
the dying person is naturally strong, the
state of his mind at the approach of death
will generally be influenced by it. Of
those who retain consciousness and intel-

---

[1] Halford, p. 75.

lect, the majority die thinking and acting
in accordance with the influences that have
been exerted upon them in previous life,
by education and example : and with those
which may be then brought to bear upon
them, towards and at its close.

More often some delirium is present.
The delirium of the dying is often of a
most interesting character, and according
to Dr. Symonds resembles dreaming more
than any form of derangement. The ideas
are derived less from present perceptions
than in insanity, and yet are more sug-
gested by external circumstances than in
the delirium of fever and phrenitis.[1] Such
delirium is generally shown in quiet talka-
tiveness, which becomes later on a low
muttering. In some the mind is occupied
on the events of childhood and early life,
but when the delirium is somewhat more

---

[1] Cyclopædia of Anatomy and Physiology,
art. Death, vol. i. p. 799.

active, the conceptions of the dying man are generally derived from subjects, which, either in his speculative pursuits or in the business of life, have principally occupied his thoughts.[1] Lord Tenterden, as he approached his end, became delirious and talked very incoherently. Afterwards he seemed to recover his composure, and raising his head from his pillow, he was heard to say in a slow and solemn tone, as when he used to conclude his summing up, in cases of great importance, "And, now, Gentlemen of the Jury, you will consider of your verdict." These were his last words; when he had uttered them, his head sunk down, and in a few minutes he expired without a groan.[2] And the last words of Dr. Armstrong were addressed to an imaginary patient, upon whom he was

---

[1] Symonds *ut supra*, p. 799.
[2] Lord Campbell, Lives of the Chief Justices of England, vol. i.

impressing the necessity of attention to the state of the digestive organs.

Instances occur, and not very rarely, where the delirium ceases, and the mind again for a time becomes clear and the sensations keen, to be followed, however, ere long by a return of delirium, or it may be of coma, or a rapid sinking of all the bodily powers and speedy death. But along with this temporary clearing of the mental powers, and in proof of its illusiveness, there are the usual signs of bodily failure—a pinching of the features, coldness of the surface, cold sweats, and a feeble rapid pulse.

Active delirium and violence are but rarely associated with the act of dying,— they are indeed scarcely compatible with it. They may pave the way to it, but when the act of dying really begins, they cease and give way to that low, rambling, muttering delirium, with which all watchers by the deathbed are so familiar.

It is especially at the stage of transition from the one to the other of these states, that we meet with that return of intelligence—that 'lightening up before death' which has impressed and surprised mankind from the earliest ages. "We have all observed," writes Sir Henry Halford, "the mind clear up in an extraordinary manner in the last hours of life, when terminated even in the ordinary course of nature ; but certainly still more remarkably when it has been cut short by disease, which had affected, for a time the intellectual faculties. We have seen it become capable of exercising a subtle judgment, when the passions which had been accustomed to bias and embarrass its decisions whilst they existed, are extinguished at the approach of death ; and when the inferences which wisdom had drawn from experience of the former behaviour of men, were now made available to a correct estimate of their future conduct, in the sense of Milton's lines —

'When old experience does attain
To something like prophetic strain.'[1]

"This is most frequently the case when the resistance of the constitution against the influence of the disease has been long protracted, or when the struggle, though short, has been very violent."[2]

"A young gentleman of family, about twenty-five years of age, took cold whilst under the influence of mercury. The disease increased daily until it was accompanied at last, by so much fever and delirium, as made it necessary to use, not only the most powerful medicines, but also personal restraint. At length, after three days of incessant exertion, during which he never slept for an instant, he ceased to rave, and was calm and collected. His perception of external objects became cor-

---

[1] On the Καῦσος of Aretæus, p. 96.
[2] Halford, On the Cautious Estimation of Symptoms, p. 17.

rect, and they no longer distressed him, and he asked pressingly if it were possible that he could live? On being answered tenderly, but not in a way calculated to deceive, that it was probable he might not, he dictated some affectionate communications to his friends abroad, recollected some claims upon his purse, 'set his house in order,' and died the following night. The reason why so unfavourable an opinion was entertained of his state, was, that the apparent amendment was not preceded by sleep, and was not accompanied by a slower pulse; two indispensable conditions — on which only a notion of real improvement could be justified. But here was merely a cessation of excitement occasioned by a diminution of power, and by a mitigated influence of the action of the heart upon the brain." This case occurred in the practice of Sir Henry Halford.[1] Another instance,

---

[1] Halford *ut supra*, p. 19.

the counterpart to that just described,
which happened to the same eminent physician, may not be out of place.

A young gentlemen, who had also been
using mercury very largely, caught cold,
and became seriously ill with fever. " His
head appeared to be affected on the fifth
day, and on the seventh, when I was first
called into consultation with another physician, who had attended him with great
care and judgment from the commencement of his illness, we found him in the
highest possible state of excitement. He
was stark naked, standing upright in bed,
his eyes flashing fire, exquisitely alive to
every movement about him, and so irascible
as not to be approached without increasing
his irritation to a degree of fury. . . .
On the eleventh day of his disease, I was
informed by my colleague, when we met, and
by the attendants, that he was become quite
calm, and seemed much better. It was
remarked, indeed, that he had said re-

peatedly, that he *should die*; that under this
conviction he had talked with great composure of his affairs; that he had mentioned
several debts which he had contracted,
and made provision for their payment,—
that he had dictated messages to his mother,
expressive of his affection, and had talked
much of a sister who had died the year
before, and whom, he said, he knew he
was about to follow immediately.   To my
questions, whether he had slept previously
to this state of quietude, and whether his
pulse had come down, it was answered, No;
he had not slept, and his pulse was quicker
than ever.   Then it was evident that this
specious improvement was unreal, that the
clearing up of his mind was a mortal sign,
 a lightening before death,' and that he
would *die forthwith*.   On entering his
room he did not notice us; his eyes were
fixed on vacancy, he was occupied entirely
within himself, and all that we could gather
from his words was some indistinct men-

tion of his sister. His hands were cold,
and his pulse immeasurably quick—he died
that night." [1]

Some pass away in sleep. In natural
healthy sleep respiration becomes slower,
the pulse weaker and less frequent, the
circulation generally feebler. The difference in these respects between the
waking and the sleeping states, is to the
dying person often the difference between
life, and death. The circulation already
reduced to the lowest ebb compatible
with life, is yet further reduced by sleep,
and with this reduction the patient dies.
These are those who 'sleep away.'
Similar to, if not identical with them, are
those to whom death comes so easily
that not a ruffle disturbs any portion of
the frame, and the most intelligent observer is unable to fix the moment when
life has fled, so easy is the parting of

---

[1] On the Καῦσος of Aretæus, p. 91.

the last link, 'when the body drops to earth and the soul rises to eternity.' It is probable that here, a mere act of dozing becomes the act of dying. In these instances as in old age, death is literally the last sleep, *uncharacterized* by any peculiarity. The general languor of the functions in the *last* waking interval, is attended with no peculiar suffering, and the last sleep commences with the usual grateful feelings of repose.[1]

The length of the interval between insensibility and the absolute cessation of existence, varies greatly from a few seconds to several hours or days. But consciousness is often retained much longer than is generally supposed, and it is difficult to determine when the external senses, and particularly that of hearing, are completely and absolutely closed.

---

[1] A. P. Wilson Philip, On Sleep and Death, 8vo, London, 1834, p. 165.

The senses of smell, taste, and touch are generally the first to fail us and disappear, while those of sight and hearing continue much longer.

Abnormal visual impressions are common when death is near at hand. In many the sight fails,—there is complaint of commencing or of actual darkness, and a desire is expressed for more light ; while more rarely, the dying one perceives a blaze of light, in the contemplation of which, or immediately afterwards, he calmly expires. " It happens not unfrequently," writes Dr. Symonds in his admirable essay on Death,[1] "that the spectra of the dying owe their origin to contemplations of future existence, and consequently that the good man's last hours are cheered with beatific visions and communion with heavenly visitors. Dreadfully contrasted

[1] Cyclopædia of Anatomy and Physiology, vol. i. p. 799.

with such visions are those which haunt the dying fancies of others." [1] The testimony of many of those who have the largest experience, and have watched *continuously and attentively* at the dying bed, is in support of Dr. Symonds' statement. If some physicians are incredulous, and place little reliance on testimony and inferences of this kind, I am inclined with Dr. Conolly [2] to attribute it, to their being seldom engaged long enough in watching by the bed-side, where the senses and thoughts naturally become

---

[1] Dr. Symonds continues, " The previous habits and conduct of the individual have sometimes been such as to incline spectators to inquire, whether in the mode of his departure from existence, he might not already be receiving retribution, just as, in other cases, celestial dreams and colloquies have seemed fitting rewards for blameless lives and religious meditation."

[2] Cyclopædia of Practical Medicine, art. Disease, vol. i. p. 629.

concentrated on the events of the sick
chamber alone. My own observation
in cases, where circumstances have made
my attendance on the dying close and
protracted, goes to corroborate the evidence
there is on these points—points which
are certainly not of a nature to be made
familiar to those, whose chief knowledge
of the dying is acquired in formal con-
sultations, or in short daily visits to the
wards of hospitals.

Hearing is, probably in most cases, the
last of our senses to leave us. " An elderly
lady had a stroke of apoplexy; she lay
motionless, and in what is called a state
of stupor, and no one doubted that she
was dying. But after the lapse of three
or four days, there were signs of amendment, and she ultimately recovered. After
her recovery she explained that she did
not believe that she had been unconscious,
or even insensible, during any part of the
attack. She knew her situation, and

heard much of what was said by those
around her. Especially she recollected
observations intimating that she would
very soon be no more, but that at the
same time she had felt satisfied that she
would recover ; that she had no power
of expressing what she felt, but that never-
theless *her feelings, instead of being painful
or in any way distressing, had been agreeable rather than otherwise.* She described
them as very peculiar—as if she were
constantly mounting upwards, and as
something very different from what she
had ever before experienced." [1]

The case of Dr. Wollaston the physician
and chemical philosopher is to the same
effect. "Some time before his life was
finally extinguished he was seen to be pale,
as if there was scarcely any circulation of
blood going on—motionless, and to all
appearance in a state of complete insensi-

1 Brodie, *ut supra*, vol. i. p. 281.

bility. Being in this condition, his friends who were watching round him, observed some motions of the hand which was not affected by the paralysis. After some time it occurred to them, that he wished to have a pencil and paper, and these having been supplied, he contrived to write some figures in arithmetical progression, which however imperfectly scrawled, were yet sufficiently legible. It was supposed that he had overheard some remarks respecting the state in which he was, and that his object was to show, that he preserved his sensibility and consciousness. Something like this occurred some hours afterwards, and immediately before he died, but the scrawl of these last moments could not be deciphered." [1]

"I have been curious," writes Sir Benjamin Brodie in commenting on these cases, " to watch the state of dying persons in

---

[1] Brodie *ut supra*, p. 182.

this respect, and I am satisfied," (and I may add, my own experience confirms Sir Benjamin Brodie's statement) " that, where an ordinary observer would not for an instant doubt that the individual is in a state of complete stupor, the mind is often active even at the very moment of death. A friend of mine, who had been for many years the excellent chaplain of a large hospital, informed me, that his still larger experience had led him to the same conclusion." [1]

Instances such as these should teach the physician and all who are about the dying, to be careful neither to say, nor do anything in the presence of the patient, which they would wish him not to hear. Their bearing on religious offices to the dying is obvious.

Sometimes, immediately preceding the very act of death, the eyelids are raised, and a look of recognition of those around

[1] Brodie, p. 182.

seems to be permitted to the dying man. Less often there is an expression of agony in the eye. "It is consolatory to know," says Sir Charles Bell, "that this does not indicate suffering, but increasing insensibility. The pupils are turned upwards and inwards. This is especially observed in those who are expiring from loss of blood. It is the strabismus patheticus orantium of Boerhaave."[1]

The nature of the disease and the mode of death exert a marked influence on the expression of face of the dying, and this is often retained by the features after death. "In some we observe the impress of the previous suffering, as in peritonitis and in cases of poisoning by irritants; in others the character is derived from a peculiar affection of some part of the respiratory apparatus; or from an affection of the

---

[1] The Anatomy and Philosophy of Expression. Fourth Edition, 1847, p. 185.

facial muscles themselves, as in tetanus and paralysis. But the condition of the mind is perhaps more often concerned in the expression than even the physical circumstances of the body. For, as some kind of intelligence is frequently retained, and strong emotions are experienced till within a few moments of dissolution, the features may be sealed by the hand of death in the last look of rapture or of misery, of benignity or of anger. Every poetical reader knows the picture of the traits of death (no less true than beautiful) drawn by the author of "The Giaour." But such observations are not confined to poets. Haller could trace in the dying countenance the smile which had been lighted by the hope of a happier existence. "Adfulgentis fugienti animæ spei non raro in moribundis signa vidi, qui serenissimo vultu non sine blando subrisu, de vita excesserunt." [1]

---

[1] Symonds, *ut supra*, p. 803.

II.

THE SYMPTOMS AND MODES OF DYING.

THE SYMPTOMS AND MODES
OF DYING.

—◦◊◦—

IT is often difficult to determine when the
act of dying really begins. Practically,
it should be dated from the moment when
the physician concludes from reliable signs,
not only that the disease under which the
patient labours is incurable by nature or art,
*but that the vital powers are already so utterly
reduced that they are beyond the possibility of
restoration.*[1]   And on these points the Father

---

[1] " At Medicus moriendi initium altius repetet,
et jam ab eo inde tempore ducet, quo signis
minime dubiis cognoverit, morbum naturæ

of Physic is perhaps still our best guide.
A sharp and pinched nose, the eyes sunk
in the orbits and hollow, the ears pale, cold,
shrunk, with their lobes inverted, and the
face pallid, livid, or black; these together
make up the celebrated *facies Hippocratica*,[1]

---

artique non tantum insuperabilem esse, sed et
sub eo vires sic perire ut reparari nequeant."
(Paradys, Oratio de Ευθαναοία naturali, p. 67).

[1] In the words of Lucretius, vi. 1,190—

"Item, ad supremum denique tempus,
Compressæ nares, nasi primoris acumen
Tenue, cavatei oculei, cava tempora; frigida pellis,
Duraque, inhorrebat tactum; frons tenta meabat:
Nec nimio rigida post artus morte jacebant."

Or, as rendered by an accomplished physician,
Dr. Mason Good—

"'Then, tow'rds the last, the nostrils close
    collaps'd ;
The nose acute; eyes hollow; temples scoop'd ;
Frigid the skin, retracted ; o'er the mouth
A ghastly grin ; the shrivell'd forehead tense ;
The limbs outstretch'd for instant death pre-
    par'd."

and show that the work of dying has commenced, and has already made some progress. They are signs of utter exhaustion in the circulation and in the muscular system, and they point to a loosening of all the bonds [1] by which being is held together in the human frame.

To these may be added the glazed half-closed eye; the dropped jaw and open mouth; the blanched, cold, and flaccid lip; cold clammy sweats on the head and neck; a hurried, shallow respiration on the one hand, or slow, stertorous breathing with rattle in the throat upon the other; a pulse irregular, *unequal*, weak, and immeasurably frequent; the patient prostrate upon his back; and sliding down towards the foot of the bed; his arms and legs extended, naked,

---

[1] "Omnia tum vero vitai claustra lababant." (Lucretius, vi. 1,151).

"Then all the powers of life were loosen'd." (Mason Good).

and tossed about in disorder; the hands
waved languidly before the face, groping
through empty air, fumbling with the
sheets, or picking at the bedclothes. These
latter symptoms come on for the most part
later in the series; they are the immediate
precursors of death, and show that that
event is near at hand.

More or fewer of these phenomena are
to be seen in most dying persons; but they
vary in number and character, in the order
of their appearance, and in their combination, according to the nature of the disease
in the course of which they occur, and of
the mode of dying to which they severally
tend.   "Although," says Sir Thomas
Watson, "all men must die, all do not die
in the same manner.   In one instance the
thread of existence is suddenly snapped, the
passage from life and apparent health perhaps to the condition of a corpse is made
in a moment: in another the process of
dissolution is slow and tedious, and we

scarcely know the precise instant in which the solemn change is complete. One man retains possession of his intellect up to his latest breath; another lies unconscious and insensible to all outward impressions for hours or days before the struggle is over." [1]

Whatever may be the remote causes of dissolution, the modes in which death is actually brought about vary remarkably, according as it begins in the heart, in the lungs, or in the brain.

Death beginning at the heart is sometimes instantaneous. Suddenly and without warning of any kind, the heart ceases to

---

[1] Lectures on the Principles and Practice of Physic. Fifth edition, 2 vols. 8vo, London, 1871, vol. i. p. 62. Sir Thomas Watson in his admirable lecture on the Different Modes of Dying, has treated the whole subject so graphically, that I shall follow him as closely as possible in what I have to adduce on this part of my subject.

4 *

beat, the individual turns pale, falls back or drops down and expires with one gasp. But oftener, death takes place slowly, there is a more or less lengthened period of exhaustion, and death occurs in the way either of syncope, or of asthenia. The phenomena which attend dying by syncope are described by Sir Thomas Watson as " paleness of the face and lips, cold sweats, dimness of vision, dilated pupils, vertigo, a slow, weak, irregular pulse, and speedy insensibility. With these symptoms are frequently conjoined nausea and even vomiting, restlessness and tossing of the limbs, transient delirium; the breathing is irregular, sighing, and, at last, gasping; and convulsions generally occur, and are once or twice repeated before the scene closes." [1]   When death occurs from asthenia or failure of contractile power in the heart, " the pulse becomes very feeble and frequent,

---

[1] Watson, p. 66.

and the muscular debility extreme, but the senses are perfect, the hearing is sometimes even painfully acute, and the intellect remains clear to the last." [1]

Death beginning at the lungs, from asphyxia or suffocation, is marked by laborious heaving of the chest, strong but ineffectual contractions of the respiratory muscles, distress about the breast; "the face at first becomes flushed and turgid, then livid and purplish, the veins of the head and neck swell, and the eyes seem to protrude from their sockets. There is vertigo, then loss of consciousness, and then convulsions." [2] The livid face and laboured breathing are accepted as evidence of severe bodily suffering, but they are only partially so, for the circulation of undecarbonized blood on which they severally depend, through the brain, in

---

[1] Watson, p. 68.
[2] *Ibid.*, p. 70.

common with other parts of the frame, first benumbs sensibility, and then abolishes it altogether. "Disturbance of respiration," says Dr. Ferriar, [1] "is often the only apparent source of uneasiness to the dying, but sensibility seems to be impaired in exact proportion to the decrease of that function."

Death beginning at the brain destroys life indirectly—by its influence on the lungs or on the heart, and so by the way of coma or of asthenia. In death by coma there is " stupor more or less profound ; the sensibility to outward impressions is destroyed, sometimes wholly and at once, much oftener gradually ; the respiration becomes slow, irregular, stertorous; all voluntary attention to the act of breathing is lost, but the instinctive motions continue. At length the chest ceases to

---

[1] On the Treatment of the Dying. Medical Histories and Reflections. Vol. iii. p. 195.

expand, the blood is no longer aërated," [1] and thenceforward precisely the same internal changes occur as in death, beginning at the lung. It is in this way that most fatal disorders of the brain produce death. When death starting from the brain acts through the heart, it occurs somewhat suddenly, and in the way of shock, as in some of the worst cases of apoplexy—the "apoplexie foudroyante," for example—or more slowly, in the way of exhaustion or asthenia, as in some cases of delirium tremens, or of phrensy—and as happened in the two cases described at pages 36 and 38.

The several modes of dying described above, are often combined in the same person, complicating the process and confusing our views of it; with the effect too, in some cases, of increasing the sufferings of the dying, but in others of lessening them. Thus coma, from implication of the brain

---

[1] Watson, p. 76.

supervening on diseases of the lung, first lessens the perception of the distress and anguish which attend them, and then extinguishes it. These mixed forms of death are seen especially in fevers.

III.

THE GENERAL AND MEDICAL TREATMENT OF THE DYING.

THE GENERAL AND MEDICAL TREATMENT OF THE DYING.

MANY of the sufferings of the death bed are not naturally or necessarily incident to the act of the dying; but are due to surrounding circumstances that admit of alteration or removal. Thus, restlessness and jactitation are often due to the weight of the bed coverings, and are at once removed by lightening them; —difficulty of breathing and gasping, increased by the heat and closeness of the chamber, are removed by the admission of fresh and cooler air, by change of posture and by pillows carefully adapted to the efficient support of the trunk of the body.

There is nothing of greater importance
in the treatment of the dying than the
right administration of nutriment. Errors
in feeding are the cause of much of the
disquietude and of many of the sufferings
that attend the dying. The sinking and
exhaustion that are in progress throughout
the system, are assumed by the attendants
to demand a free administration of food
and stimulants, forgetting that the stomach
shares in the exhaustion, and has lost its
tone, and in great part, if not wholly, its
power of digesting. Food is given too
frequently, and in quantities too large.
The dying person is induced by the
wearisome importunity of his attendants
to take food or stimulants, against which
nature and his stomach revolt. The
evident dislike and loathing with which he
submits, the difficulty he has in swallowing
it, and the urging and retching which
that act sometimes induces, ought to save
him from what is really under the cir-

cumstances an act of cruelty. " Here," to use the words of Sir Henry Holland, "we are called upon to maintain the cause of the patient, for such it truly is, against the mistaken importunities which often surround him, and which it requires much firmness in the physician to put aside."[1]   The wishes of the patient himself, when he has reached the stage of existence here contemplated, may generally be taken as a correct indication in all that relates to the administration of food and stimulants.

Food when unwisely given, accumulates in the stomach, distends and distresses it, and impedes the respiration. Under such circumstances the pit of the stomach will be found tumid and tense, dull upon percussion, and intolerant of pressure. At length some of the contents of the distended stomach regurgitate into the throat

---

[1] Medical Notes and Reflections.   Third edition, 8vo, London, 1855, p. 379.

or mouth; or there may be actual vomiting,
and this to the evident relief of the
sufferer. Hiccup is often due solely to an
overloaded and distended stomach.

Much discretion is needed in fixing on
the kind and quantity of food to be given.
Something will depend on the character of
the disease under which the patient is
sinking; and something on the length of
time he is likely to survive. If the act of
dying is likely to be protracted, as it often
is in cancer and some cases of consumption,
where death is brought about by slowly
progressive exhaustion, the food should be
supporting and in somewhat larger quantity.
I have long doubted whether strong beef-
tea and meat extracts are as a rule of much
use, or are appropriate when the act of dying
has really commenced. Milk, cream, beaten
eggs, and the farinacea are far better. They
are, too, the best vehicles for wine and
spirits; and they have less tendency than
soups to become offensive in the stomach.

Alcohol in its fermented or distilled forms is of special use in the treatment of the dying. Owing to its high diffusive power it passes readily into the blood. It stimulates the failing heart, and thus promotes the circulation through the lungs, which is one of its most valuable properties in the dying. It may perhaps increase the secretion of the gastric juice; it more probably stimulates the peristaltic movements of the stomach, and by so much, aids the digestive process, and supports the patient in the best and most natural manner. Stimulants and nutriment should as a rule be given together for they mutually influence each other.

The quantity of wine or spirit which is needed varies exceedingly, and no definite rule can be laid down on this point. They should be given in small quantities at a time and repeated at short intervals before the effects upon the heart and pulse of the previous dose have subsided.

Of wines, sherry is perhaps the most useful. Port, if preferred by the patient, may be substituted, but I have not found it, as a rule, to agree as well as sherry. Madeira from its slight acidity is specially agreeable to the palate, and is besides the most sustaining and cordial of wines. But tokay is often more acceptable than any other wine, especially to those sinking from exhausting diseases, as hemorrhage, profuse suppuration, and the like. It is best given with cream. The stimulus of these wines is longer maintained than is that of other forms of alcohol. Champagne is most refreshing and is often eagerly taken ; but its effects are evanescent and it needs repeating at shorter intervals than other wines. A teaspoonful of brandy, or of some liqueur may sometimes be advantageously added to it.

Sometimes brandy answers better than any wine, especially if the stomach is irritable and there is nausea or vomiting. As a mere stimulant it is best administered with yolk of egg and sugar, as is Sir Henry Halford's celebrated mixture—the Mistura Spiritus Vini Gallici of the Pharmacopœia. If brandy is used for its special tranquillizing influence on an irritable stomach, it may be given neat, in drachm doses, or in double that quantity in a little simple, or in one of the aërated, waters. The wish of the patient for any particular form of stimulant is almost always a correct indication for its use.

The dry and parched condition of the tongue and mouth so common in the dying, and the inextinguishable thirst that attends some forms of it, need constant attention. A spoonful of iced-water repeated frequently will be a great comfort. So, too, is a small bit of ice allowed to dissolve in the mouth—or lemonade—or weak black tea without milk, and slightly acidulated with a slice of lemon.

In the case of nutriment and stimulants as of mere diluents, it is to be understood—

supposing there is nothing to forbid—
that so long as the lips close upon them,
and an act of swallowing follows *promptly*,
they may be continued : but when liquids
seem merely to trickle down the throat,
and after a time, only to excite a faint effort
of swallowing, they should no longer be
persisted in.   The sensibility of the parts
is so diminished that the patient is insensible to the stimulus of the liquid, and
we infer *a fortiori* to the dry and parched
state of the mouth and fauces.   If, after
rubbing the lips gently with the spoon, or
with the spout of the feeding vessel, no
evident and distinct act of swallowing
follows, it is useless, and it may be cruel to
persist ; the liquid will but clog the mouth
and fauces, add to the impediment to breathing, and by so much, if any consciousness
remains, to the sufferings of the dying.

Next in value to stimulants in the treatment of the dying is opium.   It is a

tradition that John Hunter used often to exclaim, "Thank God for opium,"[1] and under no circumstances are we bound to be more thankful for it then when ministering at the bedside of the dying.   Opium is here worth all the rest of the materia medica. Its object and action must however be clearly understood.   Opium is administered to the dying, as an anodyne to relieve pain ; or as a cardiac and cordial to allay that sinking and anguish about the stomach and heart, which is so frequent in the dying, and is often worse to bear than pain, however severe.   Opium should rarely be adminis- tered to the dying as a mere hypnotic, or with a view to enforce sleep.   To do so would be to risk throwing the patient into a sleep from which he may not awake. But opium often induces sleep indirectly, and in the kindest way, by the relief of pain,[2]

---

[1] Robert Willis, M.D., On Urinary Diseases, 8vo, London, 1838, p. 100.

[2] "When there is a sudden cessation, or inter-

or sinking that had hitherto rendered sleep
impossible.

For the relief of pain in the dying where-
ever it may be situated, we have our one
trustworthy remedy in opium. Heberden
writes, " In impetu autem doloris, ubi ubi is
fuerit, opium est unicum remedium." If
judiciously and freely administered it is
equal to *most* of the emergencies in the
way of pain, that we are likely to meet
with in the dying,[1] whereas if timidly and
inadequately used, the sufferer is deprived
of the relief which it alone is capable of
affording.

The value of opium in allaying pain,
great as that is, is however second to its

---

mission, of acute pain, sleep frequently comes on
instantaneously at every such interval of ease.
The records of judicial torture furnish much
striking evidence as to these effects." (Sir Henry
Holland's Medical Notes and Reflections, p. 369.)

[1] I except hydrophobia, tetanus, &c., against
which it is almost powerless.

value in relieving the feeling of exhaustion
and sinking—of indescribable distress and
anxiety—referred to the stomach and heart,
which so often attends some part of the act
of dying.   To the practised eye, this condition is evidenced, as much by the pinched
features, pallid complexion, and *anxious
expression of face*, as by any verbal complaint of the sufferer.   Here the action of
opium is that of a cordial in the fullest
sense of the word.   "Of all cordials," says
Sydenham, "opium is the best that has
hitherto been discovered.   I had nearly
said," adds he, "that it is the only one."[1]
Under the protection of an opiate, writes
Dr. Heberden,[2] the patient's strength has

---

[1] "Praestantissimum remedium cardiacum
(unicum pene dixerim) quod in rerum natura
hactenus est repertum."   Sydenham Thomæ
Opera Omnia, edidit G. A. Greenhill, M.D.,
8vo, London, 1844, p. 175.

[2] "Vires ægri somno recreatæ sunt, atque
etiam ubi salus ejus prorsus desperata fuerit, et

been kept up, and even in hopeless cases in which the dying person is harassed by unspeakable inquietude, he may be lulled into some composure, and without dying at all sooner may be enabled to die more easily." I know of nothing in our attendance on the dying more gratifying, than to witness the improvement in face, feature, and expression, that marks the kindly action of opium under these circumstances. In an hour or thereabouts, after it has been taken, some colour returns to the face, the features lose somewhat of their sharpness, a placid expression replaces the look of anxiety, and the sufferer passes into an easy, gentle sleep, from which he awakes refreshed and comforted, and helped as it would seem, to die more easily, when his

---

angor summus cruciaverit, opium utique sollicitudinem aliquantum levavit. Mors quidem neque serius, neque citius venit, sed tamen minore cum cruciatu." (Heberden *De Ileo*.)

time arrives.   Hufeland, writing at the end
of a long professional life, did not hesitate
to declare that opium "is not only capable
of taking away the pangs of death, but it
imparts even courage and energy for
dying." [1]

---

[1] Hufeland's remarks on opium are so valuable
that I give them at length.   "Who would be a
physician without opium in attendance on cancer
or dropsy of the chest ?   How many sick has it
not saved from despair ?   For one of the great
properties of opium is, that it soothes not only
corporal pains and complaints, but affords also
to the mind a peculiar energy, elevation, and tran-
quility.   The soothing virtue manifests itself in
the most splendid manner in relieving death in
severe cases, in effecting the euthanasia, which
is a sacred duty and the highest triumph of the
physician, when it is not in his power to retain
the ties of life.   Here, it is not only capable of
taking away the pangs of death, but it imparts
even courage and energy for dying ; it promotes
in a physical way even that disposition of mind
which elevates it to heavenly regions.   A man
who had laboured for a long time under complaints of the chest and vomicas finally approached

Opium must be administered in such doses as will appease suffering and disorder, and in this respect we are to be governed

---

death. The most dreadful anguish of death with a constant danger of suffocation seized him, he got into real despair and his state was an insurmountable torment even for the persons around him. He now took half a grain of opium every hour. After three hours he became quiet, and after he had taken two grains he fell asleep, slept quietly for several hours, awoke quite cheerful, free from pain and anxiety, and at the same time so much strengthened and appeased in his mind, that he bade farewell with the greatest composure and satisfaction to his relatives, and after he had given them his blessing and many a good admonition fell again asleep and passed away while sleeping." (The Three Cardinal Means of the Art of Healing, p. 46.)

Somewhat to the same purport writes De Quincey. " Simultaneously with the conflict the pain of conflict has departed, and thenceforward the gentle process of collapsing life, no longer fretted by counter-movements slips away with holy peace into the noiseless deeps of the Infinite." (Confessions of an English Opium-Eater, p. 149.)

solely by the effect and relief afforded. The dose for an adult should be rarely less than a grain, but oftener more. "There exists," writes Sir Henry Holland, "distrust, both as to the frequency and extent of its use not warranted by facts, and injurious in many ways to our success;" [1] "its use is not to be measured timidly by tables of doses, but by fulfilment of the purpose for which it is given. A repetition of small quanti- ties will often fail, which concentrated into a single dose would safely effect all we require." [2]

---

[1] *Ut supra*, p. 516.

[2] Holland, *ut supra*, p. 518. To the same effect writes Dr. James Gregory of this remedy. "*Neque dubium est,* utcunque periculosus videatur usus talis medicamenti vix non venenati; *ægros plus fere incommodi et damni percepisse a nimis parva, quam a nimia ejus quantitate.* Medici igitur est, medicamentum adeo validum et sæpe perniciosum caute et prudenter adhibere, et in illis tantum morbis ad id confugere, qui aliquid istiusmodi plane requirunt; *ubi vero talis*

The effects of opium continue for about eight hours, and if its action is to be maintained it should be repeated at intervals of that duration or somewhat less. The dose is to be governed solely by the relief afforded. Its effects are usually limited to relief of the pain, or of the sense of sinking for which it has been given, producing no other direct effect on the system in general. " It would seem," says Sir Henry Holland, " that the medicine, expending all its specific power in quieting these disorders, loses at the time every other influence on the body. Even the

---

n.cessitas urget, oportet remedium libere et cum fiducia præscribere; tunc enim non sperare modo potest, sed fere polliceri, se effectum illum salutarem, quem cupit, per suum medicamentum esse præstiturum. *Quod si timide et nimis parce datum fuerit,* longe alium effectum habebit, et üsdem ægrotis *haud parum nocebit, quibus largius datum multum profuisset.*" (Conspectus Medicinæ Theoreticæ, § MCCXXII.)

sleep peculiar to opium appears in such instances to be wanting, or produced chiefly in effect of the release from suffering." [1]

Opium should always be given to the dying in its liquid forms—as the tincture, or the liquid extract — or as morphia, of which I know of no preparation of equal value to the solution of the bimeconate.

So long as the air passages are not obstructed by secretion, so long as there is neither lividity nor even duskiness of face, opium, if indicated, may be given in aid of the Euthanasia ; but if they are present, it is hazardous and might hasten death. Much care, too, is needed in the employment of opium, in cases where the heart is *greatly enfeebled,* and where the conditions, directly or indirectly induced by opiates, especially that of sleep, may be just enough

---

[1] *Ut supra,* p. 518.

5*

to turn the balance against it. A contracted
pupil is also a contra-indication to opium ;
it implies a state of the brain, which opium
is likely to increase rather than relieve.
And if food has been injudiciously pressed
upon the patient, so that the stomach
is distended with it, and the epigastrium
is full and tense, opium given by the
mouth is rarely found to act kindly, if at
all. If, under such circumstances, the influence of opium is needed, we should resort
to the hypodermic injection of morphia.

Professor Paradys warns us of the con-
fusion of the senses and of the mind that
sometimes follows the administration of
opium to the dying, and which to some
persons is worse to bear than the sufferings
for which it has been prescribed.[1] But this,
in my experience, has been rare, and will be

---

[1] " Audivi plus semel ægros temporarium a
narcoticis levamen enixe deprecantes, quod
sensuum obscuratione nimis care querebantur
emi." (p. 71.)

seldom observed if opium is restricted to
the cases where, as I have stated above, it
is specially called for,—namely, in relief of
pain or of severe sinking. When, however,
it does occur in these circumstances, it is
probably due, either to an idiosyncrasy on
the part of the patient, or to the inade-
quacy of the dose given, which has been
enough to confuse and stupify the senses,
but not to control the symptoms for which
it was administered.   "Si timide et nimis
parce datum fuerit," writes Dr. Gregory,[1]
" longe alium effectum habebit, et üsdem
ægrotis haud parum nocebit, quibus largius
datum multum profuisset."

Ammonia is inferior as a stimulant to
wine and brandy, which are more palatable
and preferable, while as an antispasmodic it
is very inferior to ether.   But it is useful

[1] " Conspectus   Medicinæ   Theoreticæ,"
§ MCCXXII.

where the respiration flags and the breathing is obstructed by secretion accumulating in the bronchial tubes, and the complexion is becoming dusky and livid. Five grains of the carbonate dissolved in camphor water is a good mode of administering it. Small doses of oil of turpentine are sometimes more effectual than ammonia. A drachm of the confection of turpentine rubbed up in peppermint water, is perhaps the best form of giving it.

Next in value to opium in its power of alleviating the sufferings of the dying is ether. It is specially indicated in gasping or spasmodic difficulty of breathing, whether dependent on the lungs or heart; and in flatulent distention of the stomach, attended with unavailing efforts at eructation. These two conditions are often conjoined in the dying, and then the indication for ether is the strongest. According to my experience ether is most efficient when given in combination with a few drops of sulphuric acid, as in the acid infusion of roses, or better with mint water and sugar, as in the so-called " ether punch." [1] Opium or laudanum in somewhat smaller doses than those recommended above, is often added, with great advantage to ether, when there is need of a potent antispasmodic.   In the paroxysms of severe præcordial anguish and dyspnœa that characterize many deaths from organic disease of the heart and great vessels of the chest, relief must be sought in ether and opium, or from the inhalation of the nitrite of amyl.

The fewer the drugs and the less of medicine we can do with in the treatment of the dying, the better.   Those above

---

[1] R Aq. Menthæ Viridis, f. ℥ v ſs.
  Sacchari, ℥ ſs.
  Acid. Sulphurici diluti ℳ xl.
  Sp. Ætheris comp. f. ℥ ij.
Misce ft Mistura. Pars quarta pro dose.

mentioned comprise all I have had occasion
for, and if judiciously used, they are equal
to the emergencies we are called upon to
meet. I have no wish unduly to limit
the means at our command in aid of the
Euthanasia; but when the stage of existence contemplated in these pages has once
been reached, we dismiss all thought of
cure, or of the prolongation of life, and
our efforts are limited to the relief of
certain urgent conditions, such as pain,
exhaustion, dyspnœa, spasm, and the like;
for which the remedies mentioned above
are to the full as efficient, if not really
more so, than any others as yet known.
But no medicine should be given without
a distinct—I had almost written urgent—
need for it; and the physician should form
a clear idea of the special requirements of
the case before him, and how, and by what
means they may be best accomplished. In
very many cases there is no need of medicines of any kind, and stimulants and light

nourishment *cautiously* administered, meet every requirement.   But often, and in almost all cases, at a certain period of their course, the less even of these that is given the better.   " Medici plus interdum quiete, quam movendo et agendo proficiunt," writes Livy, and there are few dying beds, where, just before the last, this precept does not find its fitting application.   " All that the dying person, then, requires is to be left alone, and allowed to die in peace."[1]

" Disturb him not—let him pass peaceably."

" The physician," writes Dr. Ferriar,[2] " will not torment his patient with unavailing attempts to stimulate the dissolving system, from the idle vanity of prolonging the flutter of the pulse for a few more vibrations: if he cannot alleviate his situation, he will protect his patient against

---

[1] Elliotson, Human Physiology, p. 1043.
[2] *Ut supra*, p. 193.

every suffering which has not been attached
to it by nature."

As the patient himself is wholly unable
to explain what is needful in his situation,
the physician is bound to act for him in
regulating the economy of the bed-
chamber. The temperature and ventila-
tion of the room—the amount of light to
be admitted — the degree of quiet to be
maintained in it — must be determined
according to the circumstances of each particular case.

When the mode of dying is by the lung,
and in the way of asphyxia, the admission
of fresh, cool air into the room seems to
conduce to the relief of dyspnœa, and
greatly to the comfort of the sufferer.

The custom of excluding daylight as far
as may be from the dying chamber, and
keeping it gloomy and dark, is in every
respect a mistake, and is to be opposed. If
there is one thing about his surroundings
which more often than any other is com-

plained of by the dying, it is of failing sight—of a darkness gathering over him; and a desire is expressed for more light.

Talking in an undertone and whispering in the presence of the dying is to be peremptorily checked. What has to be said, and the less that is the better, should be in a clear, distinct, ordinary tone, somewhat, perhaps, below the ordinary.[1]

The dying - chamber is no place for officious interference or obtrusive curiosity.

---

[1] Miss Nightingale's observations on whispered conversation in the room, or just outside the door, at p. 26 of her " Notes on Nursing," have great value and a wide application. On these points in the management of the dying chamber Professor Paradys has the following: " Sed præterea adhiberi hoc loco moderatæ sensuum externorum impulsiones utiliter possunt, quæ vividiores phantasmatum impressiones obscurent: vitari itaque nimiæ tenebræ et silentia nimis alta debent, concedi contra modica lux, permitti notæ amicorum voces, immo excitari debent lenes, placidi, animum blande demulcentes affectus." (p. 74.)

The fewer that are admitted to it the better
—the nurse, the minister of religion, the
medical attendant, and the immediate mem-
bers of the family, comprise those whose
duty and feelings entitle them to be
present.

"While the senses remain perfect, the
patient ought to direct his own conduct,
both in his devotional exercises, and in the
last interchange of affection with his
friends." [1]   He will be wise if he does
so under the experienced guidance of his
religious adviser.   "The powers of the
mind, after being forcibly exerted on these
objects, commonly sink into complete
debility, and respiration becoming weaker
every moment, the patient is rendered *apparently* insensible to everything around
him.   But the circumstances of the disease
occasion much variety in this progress." [2]

---

[1] Ferriar, p. 193.
[2] Ibid., p. 194.

Even when persons appear insensible, it is certain, as I have before remarked, that frequently they are cognisant of what is passing about them. " I have known them requested," says Dr. Elliotson, " to give a sign that they were still alive by moving a finger, or by interrupting their breath when to move a finger was impossible : and they have done so, although believed by many to have been long senseless." [1]   In many cases there is a sort of lucid interval immediately before dissolution.   This may be perceived by the looks and gestures where the patient is incapable of speaking.

When things come to the last and the act of dissolution is imminent, all noise and bustle about the dying person should be prohibited, and unless the patient should place himself in a posture evidently uneasy he should be left undisturbed.[2]   The

---

[1] Human Physiology, p. 1043.   [2] Ferriar, p. 203.

dying are often impatient of any kind of
any kind of covering.[1]   They throw off
the bed-clothes and lie with the chest bare,
the arms abroad, and the neck, arms,
and legs as much exposed as possible :—Ubi
supinus æger jacet, porrectis manibus et
cruribus, writes Celsus—ubi brachia et crura
nudat et inæqualiter dispergit.   " These
actions," writes Dr. Symonds,[2] " we believe
to be prompted by instinct, in order that
neither covering nor even contact with the
rest of the body may prevent the operation
of the air on the skin.   There are actions

---

[1] " Nihil adeo posses quoiquam leve tenueque
       membris
    Vortere in utilitatem."
                     (Lucretius vi. 1169.)

        " Nor would once endure
The lightest vest thrown loosely o'er the limbs."
                     (Mason Good, p. 595.)

[2] " Cyclopædia of Anatomy and Physiology,"
vol. i. p. 802.

and re-actions between the air and the blood
in the skin similar to those which occur in
the lungs, and these are in aid of them."
Such automatic actions ought not to be
interfered with, unless the patient has got
into a position evidently distressing to himself, or except so far as decency requires
when there is any approach to unseemly
exposure.

Exclamations of grief, and the crowding
of the family round the bed, only serve to
harass the dying man, writes Ferriar, who
adds, " The common practice of plying him
with liquors of different kinds, and of forcing
them into his mouth when he cannot swal-
low, should be totally abstained from."
But to this error I have already referred.

It was a custom in the Middle Ages to strip
the dying, drag them from their beds, and
lay them on ashes or on mattrasses of straw
or hair upon the floor. It was then wholly
or in part a penitential act, and the influence
of this custom has, perhaps, not yet wholly

ceased. "It is," says Dr. Ferriar,[1] "a
prevalent opinion among nurses and servants that a person whose death is lingering
cannot quit life while he remains on a common bed, and that it is necessary to drag the
bed away and place him on the mattrass.
This piece of cruelty is often practised when
the attendants are left to themselves. A
still more hazardous practice has been very
prevalent in France and Germany, and I
am afraid is not unknown in this country.
When the patient is supposed by the nurses
to be nearly in a dying state, they withdraw
the pillows and bolster from beneath the
head, sometimes with such violence as to
throw the head back and to add greatly
to the difficulty of respiration. As the
avowed motive for this barbarity is a desire
to put the patient out of pain—that is,
to put him to death—it is incumbent on
his friends to preserve him from the hands

---

[1] P. 200.

of those executioners. Perhaps a more deplorable condition can scarcely be conceived than that of being transferred from the soothing care of relations and friends, to the officious folly or rugged indifference of servants." One would hope that such cruelty is a thing of the past. My own experience forty years since as a dispensary physician in the eastern parts of the metropolis, led me to conclude that it was not *then* and *there* wholly unknown or unpractised. What it may be in remote rural districts, where the class of old, ignorant and prejudiced nurses still exist, I have no means of knowing. "This is a state of suffering," adds Dr. Ferriar, "to which we are all exposed, and if it were unavoidable, I should be far from desiring to unveil so afflicting a prospect. But the means of prevention are so easy, that I cannot forbear to solicit the public attention to them."[1]

---

[1] P. 203.

In the intelligent trained nurses of the present day, we have the best security against such barbarity ; and when they are absent, in the presence in the dying chamber, of the relations or nearest friends until all is over.

In cases of sudden death from disease of the heart, there is neither occasion nor time for medical treatment of any sort. Death is instantaneous and without warning. Where death beginning at the heart takes place by way of syncope, fresh air and stimulants cautiously given are the best resources. Wine or brandy, with egg or other light nutriment, are appropriate, When death is taking place in the slower way of exhaustion, a like treatment is to be pursued. In the earlier stage, small quantities of soup, or beef tea may be given, but when death is near they are best omitted. It is in these cases that madeira and tokay answer so well. In all cases of dying by failure of the heart's action, the posture of the patient should be carefully adjusted—the head should be low rather than raised, and it and the shoulders supported on firm pillows. Any approach to the erect or sitting posture is as a rule to be avoided. Its tendency is to occasion fainting and death.

In death from the lungs or by asphyxia the struggle is often protracted, and accompanied by all those marks of suffering which the imagination associates with the closing scene of life. Doubtless in the earlier stages of it, there is real suffering, but happily this is rarely of long duration, for the circulation of venous blood ensues, and deadens sensibility and pain. The respirations in this mode of death become laborious and heaving, the expression of countenance distressed and anxious. But soon the face becomes tumid and dusky, the lips livid, and with the circulation of undecarbonized blood, which these symptoms imply, the anxious expression of face sub-

6

sides, and there ensues a slowly increasing
benumbing of sensation, and a corresponding
diminution of suffering. The breathing
then becomes irregular and laborious, and
the heavings of the chest convulsive; but
these movements are automatic, and in-
dependent alike of sensation and of the
will. They soon pass into coma, stertor,
rattle in the windpipe, and death. Stertorous breathing is in great measure due to
affection of the brain or medulla, either
primary or secondary. The latter is the
condition we are here contemplating.
Stertor seems to be due to a falling back of
the base of the tongue into the pharynx,
and to the obstruction to respiration thence
induced; and is increased by the prone position on the back, into which such patients
naturally fall. It may be relieved by placing
the person on one side, and supporting him
in that position by well-arranged pillows.
The tongue then drops to the side of the
pharynx and mouth, and leaves room for

the ingoing air. Dr. Bowles, of Folkestone, to whom we owe the knowledge of these facts, warns us, that care should be taken to keep the neck rather straight, as, if the chin be brought too near the sternum, the thyroid cartilage presses upwards and backwards, and again pushes the base of the tongue, toward the back of the pharynx. Nothing can be done, indeed nothing is needed, but regulation of the posture, when coma is established.   The head is to be supported on a firm pillow, or bolster, and slightly raised, but not so much as to increase the tendency to slide downwards in the bed.   Whatever position of the body is found to lessen the stertor, and ease the breathing should be maintained.

In the earlier stages of the process above described—in the condition which precedes and passes into coma—a carefully adjusted posture of the patient, in which he is propped up at an angle of not less than forty-five degrees, and often at one of much more, and due support is given to the trunk of the body by pillows—will do more than anything else in relief of embarrassed and laboured breathing. "The object is to support with the pillows, the back *below* the breathing apparatus, to allow the shoulders room to fall back, and to support the head, without throwing it forward."[1] The suffering of dying patients, says Miss Nightingale, is immensely increased by neglect of these points. If secretions have accumulated in the air passages, ammonia or turpentine may be administered. Should the breathing be gasping and spasmodic, ether, with or without opium, should be tried. When duskiness and lividity of the face have come on, we can do but little—when deep coma and stertorous breathing, nothing— but adjust the posture of the patient to

[1] Miss Nightingale's Notes on Nursing, p. 47.

the more pressing requirements of the case.

When the heart or great vessels of the chest are the seat of the disease, and the circulation through the lungs is becoming seriously embarrassed by it, there are often paroxysms of great suffering.   The patient is agonized by a sense of instant suffocation, and sits in or out of bed, with the head bent forward, resting on a table or other support, and expecting dissolution every moment. Here ether and opium is our best resource; or the nitrite of amyl, the cautious inhalation of which has in some instances given marked relief.

When death, commencing at the brain, destroys life through the lung, and in the way of coma, as it usually does, the treatment is the same as in the coma that occurs late in the series of events which mark death by asphyxia.   When, on the other hand, death, beginning at the brain, destroys life through the heart and by way of exhaus-

tion, the treatment is the same as above
described for those dying primarily from
the heart and in the way of asthenia.

When the face of the dying person is
flushed, the head hot, and the carotid
arteries beating forcibly, the head is to be
raised and supported on firm pillows, and
ice or a cold spirit lotion applied to it.

In some delicate and highly sensitive
persons, a kind of struggle is sometimes
excited when the respiration becomes very
difficult.[1]   Dr. Ferriar says he has known
this effort proceed so far, that the patient
a very few minutes before death, has started
out of bed, and stood erect for a moment.
He ascribed it to apprehension and alarm,
and adds: " Those who resign themselves
quietly to their feelings seem to fare best."
This is probably true, but the sufferer

---

[1] Ferriar, *ut supra,* p. 196.

needs whatever relief art can supply; and ether and opium is the most likely to give it.

Hiccup is somewhat alleviated by a sinapism to the epigastrium, and a spoonful of aniseed water swallowed slowly. But if it is severe, shaking the patient, and so adding greatly to his distress, we must rely on opium given internally, and its application externally to the pit of the stomach. If hiccup seems to be due, as it often is, to an overloaded and distended stomach, and the influence of opium is needed, the hypodermic injection of morphia is to be preferred.

Inquietude and restlessness, especially in the half-conscious dying person, is often due to a distended bladder, and is at once quieted by the catheter. In others, it is due to the weight of the bedclothes, and is relieved by lightening them.

Coldness of the feet is best met by a foot warmer; and not by thick, heavy bed clothing, which distresses the sufferer and gives rise to inquietude and restlessness. "Weak patients," says Miss Nightingale, and the dying as much or more so than others, "are invariably distressed by a great weight of bedclothes." Light Whitney blankets should alone be used for coverings under such circumstances. But I am not sure that coldness of the extremities does always add to the sufferings of the dying, or needs the consideration usually given to it. The diminished circulation on which it depends is attended, for the most part, in the dying by proportionate loss of sensibility; and besides it is especially when the feet and legs are cold, sodden, and dank, that we observe that impatience of any covering upon them—that tossing about and exposure of them to the air—which I have before described.

Death from old age—the natural termination of life, and the simplest form of death that can occur, creeps on by slow and almost imperceptible degrees. It is characterised by a gradual and proportionate decay of all the functions and organs of the body, and as a rule presents no symptoms that call for special treatment. It is only where the normal course of decay is disturbed by supervening disorder, or disease of an important organ, or by surrounding circumstances, that suffering of any kind attends it. Good nursing, and the due administration of light food and stimulants, comprise all that is needed. The approaches to death are so gentle, and the act of dying so easy, that nature herself provides a perfect euthanasia.

THE END.

UNWIN BROTHERS, PRINTERS, CHILWORTH AND LONDON.

www.ingramcontent.com/pod-product-compliance
Lightning Source LLC
Chambersburg PA
CBHW021825190326
41518CB00007B/751